D1511446

Living on the Edge

Icy Antarctic Waters

WENDY PFEFFER

BENCHMARK BOOKS

MARSHALL CAVENDISH
NEW YORK

Contents

Weddell seals, minke whales, and emperor penguins spend their whole lives in this frozen land.

Life in Icy Waters

Penguins, whales, and seals swim in the Antarctic's icy waters and face fearsome challenges. Emperor penguins go for months without food. Minke (MING-kee) whales swim under ice for miles without air. Weddell seals are born on ice floes. In any of these situations people would die, but these hardy animals survive.

The Antarctic surrounds the South Pole. Antarctica is the iciest, windiest, and coldest continent on Earth. Parts of the continent lie buried under ice nearly 3 miles (5 kilometers) thick. There is no vegetation to break the wind's fury across its 5.5 million square miles (14 km²). Antarctica holds the record for the coldest temperature on Earth, –96 degrees Fahrenheit (–71°C). The temperature sometimes stays below freezing even in summer.

Many countries have sent explorers and researchers to learn about this vast, isolated continent. In winter the climate is so frigid, eyelashes become coated with ice. The sweat inside clothes freezes, and eyelids may freeze shut.

The lack of food and shelter and the presence of blustery, icy winds make it difficult for anything to stay alive. Only a handful of scientists live year-round at research stations clustered along Antarctica's coasts. The largest land animal is a wingless midge, only .5 inch (1.2 centimeters) long. Most life exists on the sea ice or in the water.

There, stormy seas and temperature changes stir up nutrients from deep down. This lets microscopic plants grow. Shrimplike animals called krill drift in and eat the plants. Krill provides food for the penguins, whales, seals, and fish that live in the frigid waters.

Scientists in a research tent

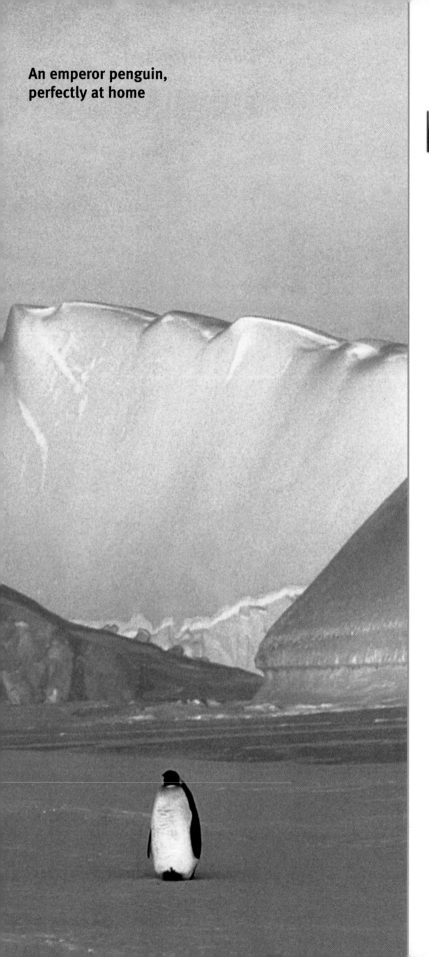

An emperor penguin, perfectly at home

Emperor Penguins

Of the seventeen types of penguins, the emperor penguin is the only one that has adapted to living all year in the bitter cold. When most Antarctic birds head north for warmer places, emperors travel south where it's colder. These winter nesters swim or hitch rides on ice floes until they find solid sea ice. There, they gather in colonies, mate, and raise their young in the harshest weather on Earth.

On the ice emperors don't need camouflage. They have no predators there. But in the water leopard seals and killer whales swimming underneath look up and see a white belly. It looks like the sky. Predators above them look down on a black back. It matches the darkness of the sea.

Emperor penguins, sea birds that don't fly, are superbly adapted to living in icy water. Short, oily feathers with tips that overlap like roof shingles waterproof them. Tightly packed and four layers deep, these feathers also keep them warm. Eighty feathers fit in a space the size of a postage stamp. Under the feathers, dense woolly down acts like the thermal underwear you might wear to keep warm. And a thick layer of fat under their skin, called blubber, insulates them.

Sometimes they dive as deep as 900 feet (274 meters) to feed. That's almost as deep as three football fields are long. A person's lungs would be crushed if he or she dived that deep. Emperors dive down for squid and fish. Their excellent underwater vision

Tightly packed feathers, woolly down, and blubber protect penguins from the freezing cold.

Try as it might, an emperor penguin cannot fly. But it would easily beat any other bird in a swimming contest.

helps them to find food. Slippery fish are hard to grip, but an emperor's spiny tongue and powerful jaws grasp them with ease.

Emperors swim by moving their oarlike flippers. This develops large breast muscles. Less than 4 feet tall (1.2 m) and weighing about 66 pounds (30 kg), an emperor's solid, chunky body helps it dive. No bird that flies could carry this weight.

But emperors seem to fly in water. Their streamlined torpedo-shaped bodies let them slice through icy waters at 25 miles (40 km) an hour—faster than any other water bird. Powerful flippers help them balance. Their webbed feet and triangular tails, made of stiff, pointed feathers, steer them in water and support them on ice.

Emperor penguins spend a lot of time underwater, out of the chilling wind. But just like you, they can't stay submerged all the time. To breathe, they leap out of the water like porpoises and gulp a breath of air. Emperors escape predators by swimming full speed ahead, then popping out of the water like slippery watermelon seeds. They don't lose speed, direction, or rhythm when landing on

Emperor penguins must first reach great speeds underwater before leaping onto sea ice.

To keep from freezing in winter winds, emperor penguins huddle.

the sea ice. Fortunately, tough feathers, stretchy skin, and thick blubber protect them when they hit its hard, cold surface.

An emperor waddling clumsily over the ice looks much like a baby taking its first steps. Sometimes an emperor flops down on its belly and slides. Emperors can toboggan faster on their bellies than people can run on snow.

On solid sea ice in the dark of winter, emperors form breeding colonies. One colony might have twenty thousand pairs. And it's noisy! Each emperor has a unique voice. Males and females recognize each other by their calls.

Emperor penguins must breed in winter so their young can be on their own by summer. The female lays one egg, the size of a softball. On the ice there are no nesting materials. So the male scoops the egg from his mate's

A group of emperor penguins tobogganing

An emperor chick must remain inside its parent's brood pouch to keep from freezing to death.

feet and into a brood pouch just above his feet. He drapes a warm fold of his feathered skin over the egg to keep out the cold.

Now the hungry female travels to open water. She spends two dark months feeding in the sea. Meanwhile, her mate protects their egg. With no food available he lives on his body fat until his mate returns.

Emperors must cooperate in order to survive hurricane-force winds. To keep from freezing, thousands of males huddle. With their bodies pressed close together, the birds shield one another. Each takes a turn in the center of the huddle protected from the cold. Then each one shuffles slowly to take a turn on the windy, chilly outer edge.

When the temperature dips, the male emperors close their feathers and hold their flippers close by their sides. They also stay warm by shuffling. But when an emperor moves around, he still carries the egg. He must be careful, because if he drops the egg, it will freeze.

After sixty-five days the egg hatches. The male regurgitates a milky liquid from his throat to feed the chick. Meanwhile, the female returns from the sea. The parents greet each other excitedly with calls and head-bobbing. Then the female moves the chick from the male's warm brood pouch to hers. This must be done quickly. Two minutes on the ice and the chick will die.

By this time, the male has not eaten for nearly 120 days, a record for any bird, so he travels to the sea and eats for three or four weeks. The female regurgitates fish and feeds their chick. She puts her mouth over the chick's

A female penguin closes her mouth over her chick's mouth when it feeds.

mouth so other birds, such as skuas, can't steal the food.

When the sea ice melts, the colonies find themselves closer to the open water. In this semidark season both parents catch fish for their fast-growing chick. At two months the chicks are too big for the pouch and must huddle together to stay warm. By five months the chicks drift north on ice floes and molt. The emperor chicks shed their baby down and grow adult waterproof feathers. Now they can jump into the sea to find their own food. Some die young but the ones that survive may live more than twenty years.

Five-month-old emperor
penguin chicks

A minke whale depends on its blubber to keep warm in freezing cold water.

Minke Whales

Like penguins, minke whales live in the Antarctic. And, like penguins, they are warm-blooded. Their body temperature remains warm no matter how icy their surroundings. Both minke whales and penguins have a layer of blubber to keep them warm. Just as you put on more clothes in winter than in summer, whales living in near-freezing water have more blubber than whales in warm water.

Unlike penguins, which lay eggs, minke whales are mammals, which give birth to live young. Babies are born in winter. Even though the calf can swim right after birth, its mother pushes it to the surface. The calf fills its lungs for the first time. For about six months it nurses underwater. A mother won't leave a wounded calf, even if it means she may be captured herself.

Minkes are baleen whales. They have no teeth. Instead, slender, flexible baleen plates hang in their mouths to filter krill from the water. The plates are made of a material much like your fingernail. The minke sweeps through the water with its mouth open and throat expanded. It takes in a huge mouthful of seawater and krill and lets the plates act like a sieve, separating the krill from the water. The whale swallows thick "krill soup." It eats about 880,000 krill a day.

Minke whales grow to be 30 feet (9 m) long and weigh 10 tons (9080 kg), as much as two elephants. A pointed snout and streamlined body help make these massive mammals fast swimmers. They can keep up with a ship traveling 10 miles (16 km) per hour.

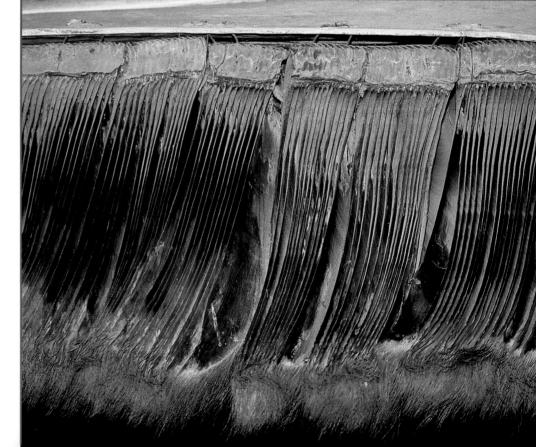

Baleen plates filter krill from seawater.

Minke whales can reach speeds of up to ten miles (16 km) per hour.

Minke whales hold their breath for a long time underwater and then breathe out hard, or blow, when they surface.

When a minke whale dives underwater, its body adjusts to the pressure. But it will drown if it stays underwater too long. Unlike fish, which breathe underwater, a whale holds its breath. The air in its lungs becomes warm and full of water vapor. When the whale comes to the surface and blows, hot air hits cold air and forms a cloudy vapor, just as your breath does on a cold day.

The edge of the solid ice is a dangerous place for sea creatures. Huge patches of pack ice come and go with the wind, sometimes trapping minke whales. They look for openings in the ice, called polynyas. To reach these, they must swim under the ice without air, sometimes for miles. When they find an opening, they surface and take deep breaths. If a minke whale can't find its way to the open sea, it might spend the whole winter trapped in a polynya, an easy target for a passing predator.

A minke whale swims in an opening in the ice, called a polynya.

A minke whale's main enemies are killer whales and whalers. The minke is the only baleen whale that continues to be hunted. In the past, minke whales were too small to bother with. Today, even though the International Whaling Commission has banned all commercial whaling, they are still being whaled in the Antarctic under scientific permits.

Their other enemy, the killer whale, can devour a baleen whale twice its size. If a minke whale escapes its enemies, it may live forty to fifty years.

If it can outswim its enemies, a minke whale can live forty or fifty years.

A Weddell seal can stretch out and relax in below-freezing temperatures because its fur is so thick.

Weddell Seals

The Weddell seal breeds closer to the South Pole than any other mammal. And it's not an easy place to survive. The Weddells, named after the explorer James Weddell, spend the whole winter in the water under the Antarctic ice sheet.

Under the thick ice it's dark even in the summer. Weddell seals communicate with one another with sounds that bounce off the ice and travel underwater. To get air, they gnaw holes in the ice, then push their snouts out to take a breath. Some sleep hanging in the icy water with only the tips of their noses poking above the surface. Others float against the underside of the ice until they need to breathe. Since ice forms constantly, Weddells use specially adapted front teeth to grind the ice around their holes. If they don't the ice will freeze over, and the seal will drown.

To get air, Weddell seals either find gaps in the pack ice or grind new openings with their teeth.

Each male claims an area around his hole in the ice. Any female that uses a male's ice hole mates only with that male. In spring when the pack ice is still more than 6 feet (2 m) thick, the female grinds a crack in the ice with

her teeth and makes a hole big enough to climb through. Then she swims to solid ice to give birth.

When a Weddell pup is born, it moves from its mother's warm body onto the ice. It must survive a drop in temperature of more than 100 degrees Fahrenheit (37.8 °C). The mother shelters her baby from the raging wind. The pup looks like a bag of skin and bones flopping around on the ice, but in just

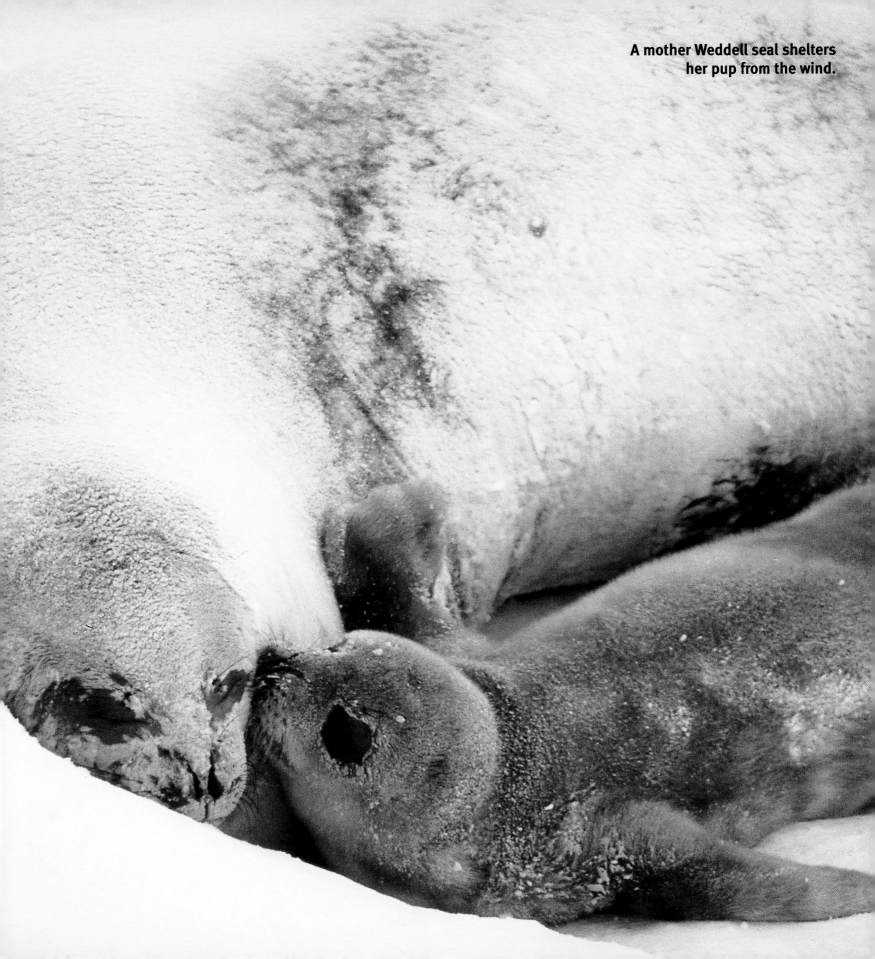

A mother Weddell seal shelters her pup from the wind.

six weeks it gains 200 pounds (91 kg). The mother, a lumbering lump of blubber, loses 300 pounds (136 kg) providing milk for her pup.

By summer, when other seals arrive, the Weddell pups can already swim. Hopefully, they can escape the leopard seals that swim in from the ocean late in the year and hunt at the edge of the sea ice.

A Weddell seal is as comfortable on land as it is deep underwater.

Weddell seals dive down 2,000 feet (600 m) for fish and squid. A human couldn't stand the pressure of such deep water, but the Weddell seal can. Before a dive it exhales, which temporarily collapses the lungs. This leaves the seal with little oxygen, but that's no problem. The seal's blood and muscles hold most of the oxygen. And the seal doesn't need as much anyway, since its heart

Just another day in the freezing Antarctic

rate slows down in a dive. The Weddell seal's body has adapted in many ways over the years so that it's able to exist under the Antarctic ice pack.

Emperor penguins, minke whales, and Weddell seals live in the coldest and windiest place on Earth. They have all adapted to the icy waters, freezing weather, blinding blizzards, and deep-sea dangers of their harsh habitat.

Other Animals Adapt and Survive

Crabeater seals don't eat crabs. They strain krill from the Antarctic's icy waters through the spaces in their triangular-shaped teeth, much the same as the baleen whales feed.

Antarctic starfish move so slowly and live at such a slow pace that some live thirty-nine years, many times longer than their cousins in warmer water.

Antarctic ice fish are cold-blooded animals living in icy-cold water. So why don't they freeze? These strange, ghostlike fish have sugars and proteins in their blood that act like antifreeze. This mixture keeps their blood from freezing.

Killer whales are amazing hunters. When they detect a seal or other prey on an ice floe, the entire pod leaps from the water. As they crash down, they make a wave that sweeps the seal off the ice, an easy catch for the whales. They also swim underwater, butt the ice, and break it with their heads, forcing the seals into the water.

GLOSSARY

ADAPT　　　　　to adjust to changes in one's surroundings

BROOD POUCH　　a pocket of bare skin, crisscrossed with blood vessels, where an egg is placed for protection; heat from the parents' blood keeps the egg warm; a flap of skin covers the egg and keeps out the cold.

CAMOUFLAGE　　coloring or body shape that makes an animal hard to see in its natural surroundings

COLD-BLOODED　having blood that becomes colder or warmer as the surrounding air or water temperature changes

COLONY　　　　a group of animals living together

INSULATE　　　to cover with a material that keeps heat or sound from escaping

MAMMAL　　　　a warm-blooded animal with glands in the female that produce milk to feed its young

MOLT　　　　　to shed skin, fur, or feathers before getting a new growth

NUTRIENT　　　a substance that promotes the growth of living things

PACK ICE　　　ice formed when slabs of sea ice freeze together

POD	a group of whales
POLYNYA	an area of open water surrounded by ice
PREDATOR	an animal that kills and eats another animal
PREY	an animal that is food for another animal
REGURGITATE	to bring partly digested food from the stomach back to the mouth
SEA ICE	ice that forms from seawater next to land

FIND OUT MORE

Billings, Henry. *Antarctica*. Danbury, CT: Children's Press, 1994.

Bunting, Eve. *The Sea World Book of Whales*. San Diego, CA: Harcourt Brace Jovanovich, 1987.

Fothergill, Alastair. *A Natural History of the Antarctic: Life in the Freezer*. New York: Sterling Publishing, 1995.

Gilbreath, Alice. *The Arctic and Antarctica: Roof and Floor of the World*. Minneapolis, MN: Dillon Press, 1988.

Pringle, Laurence. *Antarctica: The Last Unspoiled Continent*. New York: Simon & Schuster, 1992.

Sayre, April Pulley. *Antarctica*. Brookfield, CT: Twenty-First Century Books, 1998.

Web Sites

http://www.pbs.org/wnet/nature/antarctica/index.html
(Antarctica: The End of the Earth)

http://library.thinkquest.org/28779
(Antarctica: The Continent of Wonder)

INDEX

Page numbers for illustrations are in **boldface**.

ABOUT THE AUTHOR

Wendy Pfeffer, an award-winning author of fiction and nonfiction books, enjoyed an early career as a first grade teacher. Now a full-time writer, she visits schools, where she makes presentations and conducts writing workshops. She lives in Pennington, New Jersey, with her husband, Tom.

With thanks to Dr. Dan Wharton, director of the Central Park Wildlife Center, Wildlife Conservation Society, for his expert review of this manuscript.

Benchmark Books
Marshall Cavendish
99 White Plains Road
Tarrytown, NY 10591-9001

www.marshallcavendish.com

Text copyright © 2003 by Wendy Pfeffer
Map by Sonia Chaghatzbanian
Map copyright © 2003 by Marshall Cavendish Corporation

Pfeffer, Wendy, 1929-
Icy Antarctic waters / by Wendy Pfeffer.
p. cm. — (Living on the edge)
Includes bibliographical references and index.

Summary: Provides information on the hardy animals that make
the Antarctic's icy waters their home.

ISBN 0-7614-1438-X

1. Zoology—Antarctica—Juvenile literature. [1. Zoology—Antarctica]
I. Title. II. Living on the edge (New York, NY)
QL106 .P54 2002
591.77'09167—dc21

2001007291

Photo research by Candlepants Incorporated

Cover photo: Phots Researchers/Art Wolfe

The photgraphs in this book are used by permission and through the courtesy of: *Animals Animals*: G.L.
Kooyman, 1, 9, (top), 11, 17; Johnny Johnson, 9 (bottom), 16, 28 (bottom); Osborne, B. OSF, 10; Allen, D.
OSF, 14, 25, 26, 27, 32, 33; Stefano Nicolini, 22; Price, R. OSF, 30-31. *Photo Researchers*: Tim Davis, 4-5;
David Vaughn/Science Photo Library, 7; G. Robertson/Jacana, 8; Tim Davis, 12; Francois Gohier, 18, 20, 23,
34-35; Gilbert S. Grant, 19; Robert W. Hernandez, 28 (top); Art Wolf, back cover. *Minden Pictures*: 13. *Norbert
Wu*, www.norbertwu.com: 21. *Corbis*: Peter Johnson, 24.

Printed in Hong Kong
1 3 5 6 4 2